Tess
and the MYSTERY Rock
FIELD NOTEBOOK

by Melinda Bagby

Written by: Melinda Bagby
Illustrations: Jane Gross

Tess and the Mystery Rock Field Notebook

ISBN 978-1-941181-61-4

Published by Gravitas Publications Inc.
www.gravitaspublications.com
www.realscience4kids.com

GRAVITAS
PUBLICATIONS

INTRODUCTION TO YOUR FIELD NOTEBOOK

Welcome to your field notebook! This is the type of notebook you and other scientists take with you into "the field," wherever you go. Now you can write down your observations, thoughts, questions, ideas, and discoveries whenever you think about or see something interesting.

In this field notebook you will find the *Thought Stop* sections from the storybook as well as real science experiments to perform. You will explore sections like *Think About It*, where you can write about what you think might happen, and *Observe It*, where you can record exactly what you see. Sometimes what you think might happen is very different from what you observe actually happening—and that's OK! Discovery is part of doing real science. Another important part of doing science is using your observations to draw conclusions about why you might have observed what you did. It's fun to draw conclusions and think about why things work the way they do.

Scientists are people who are curious and ask lots of questions. Doing science involves looking for answers to these questions by doing research to find out what others have discovered and by dong experiments to make your own discoveries. So, no answers are provided in this notebook, and there are no "right" answers. See what you can think of and discover for yourself!

Feel free to write anywhere in this notebook. It's *your* notebook! Answer the questions, record your observations, and write notes sideways in the margins or on the top or bottom of the page. Ask yourself a question like, "What would happen if...?" and then come up with your own experiment to find out. You can add drawings too and stick in more pages. This is your science notebook and you get to decide what goes in it. And when you've filled this one up, you can start a new one.

Your field notebook will help you have fun with real science!

Tess and the Mystery Rock FIELD NOTEBOOK

TESS and the MYSTERY Rock
FIELD NOTEBOOK

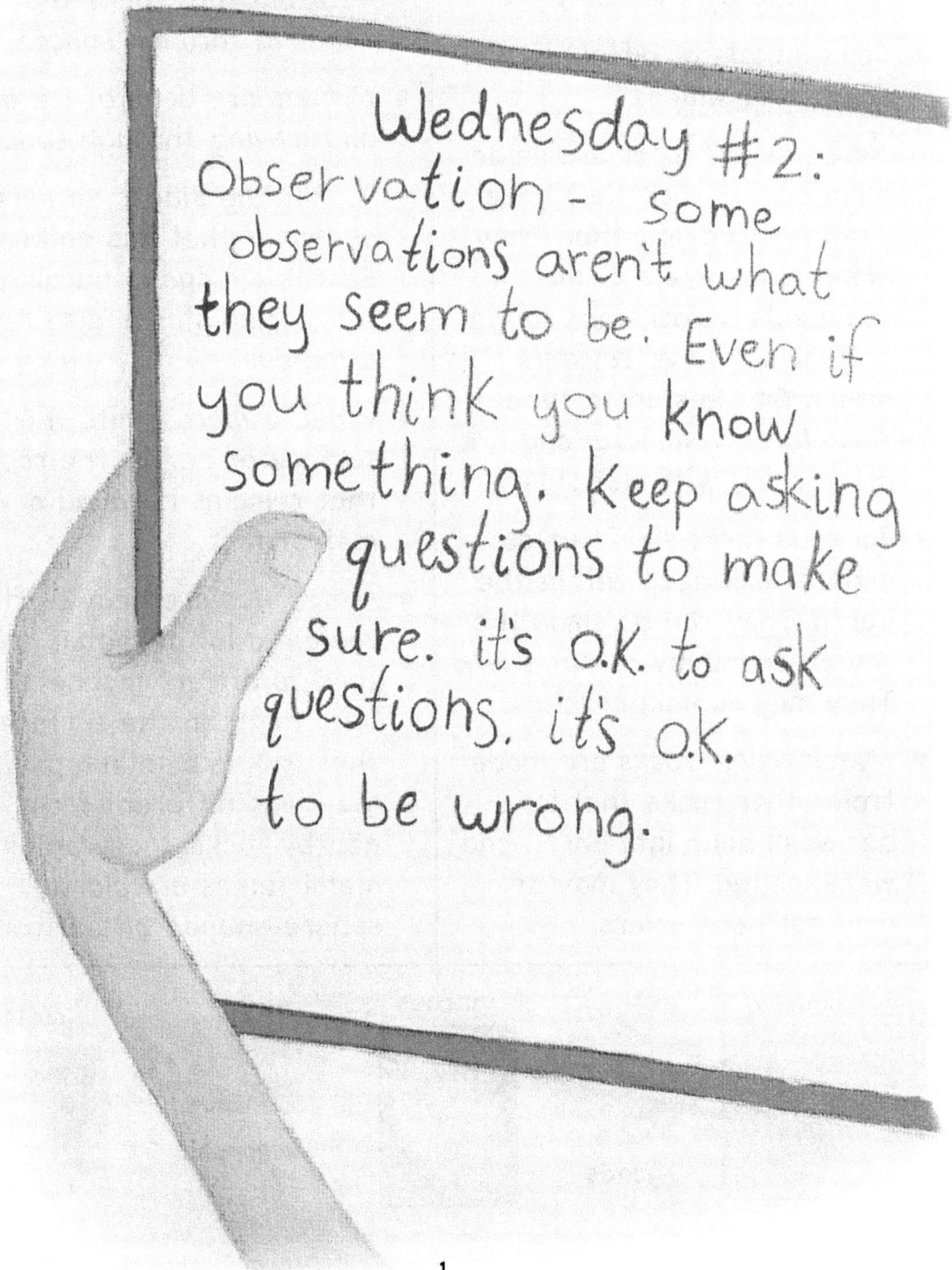

Wednesday #2:
Observation - some
observations aren't what
they seem to be. Even if
you think you know
something. Keep asking
questions to make
sure. it's O.K. to ask
questions. it's o.k.
to be wrong.

Earth Rocks   vs.   Space Rocks

Earth Rocks...

- Are made by the Earth.

- Come in lots of different colors and shapes.

- *Sedimentary rocks* are made from bits of rock that have been pressed together. They often have layers. Some sedimentary rocks are round or oval and have *crystals* inside. Crystals are a type of *rock-forming mineral* and are often shiny with flat sides.

- *Igneous rocks* start out as melted rock deep within the Earth. They can be smooth or rough, and shiny or not shiny. They may contain crystals.

- *Metamorphic rocks* are made from other rocks that long ago sank deep into Earth and were melted. They may or may not have layers.

Space Rocks...

- *Asteroids* are rocks that are zooming through space.

- *Comets* are balls of ice and dirt moving through space.

- A *shooting star* is an *asteroid* or *comet* that has entered Earth's air and is burning up. A shooting star is also called a *meteor*.

- When a *meteor* hits the surface of Earth, the rock that remains is called a *meteorite*.

- *Meteorites* are heavy, often contain a lot of metal, and tend to attract magnets. May have dents in the surface that look like thumb prints. May look different from nearby rocks. May contain metal flakes or colorful, sphere-shaped bits of rock.

sedimentary

igneous

geode

metamorphic

meteorite

Photo credits: Geode and Meteorite, by H. Raab

Think About It

❶ How would you describe a meteorite to your friend?

_____

_____

_____

❷ How do you think a meteorite is different from other kinds of rock?

_____

_____

_____

❸ List the things that you think would help you tell whether or not a rock you find is a meteorite.

_____

_____

_____

❹ Do you think Tess found a meteorite? Why or why not?

_____

_____

_____

_____

How To Be a Scientist

When scientists want to find out more about something, they follow a certain set of steps. These steps are called the *scientific method*.

❶ The *first step* is to make *good observations*. A good observation happens when you look at something in detail.

❷ The *second step* in the scientific method is to ask a question about what has been observed and then turn that question into a *hypothesis*. A hypothesis is a statement about something. For example, you might wonder, "Which is heavier, an elephant or a kangaroo?" Turning that question into a hypothesis, it becomes, "An elephant is heavier than a kangaroo."

❸ The *third step* in the scientific method is to do an *experiment* to prove or disprove the hypothesis. In the elephant-kangaroo example, the scientist would weigh the elephant and the kangaroo one at a time.

❹ The *fourth step* of the scientific method is *collecting results*. In the elephant-kangaroo example, to collect the results, the scientist would write down the weights of the two animals.

❺ The *fifth and last step* of the scientific method is to *draw a conclusion*. What did the information show? Was the elephant heavier than the kangaroo or was the kangaroo heavier than the elephant? The conclusion comes from the *results* of the experiment. Based on the results, the scientist draws a conclusion and shows that the hypothesis has been either proven or disproven.

Think About It!

Imagine you are with Tess and her friends. What do you think Tess and her friends will find inside the rock? Do you think it will be solid? Do you think the inside will look like the outside? Do you think it will be full of colored rocks? Do you think it will be full of water? Do you think it might have clay on the inside? Do you think something else might be inside the rock?

On the next page write down your *hypothesis*. Your hypothesis is your guess, or prediction, about what you think Tess and her friends will discover. Base your hypothesis on the information you have about Tess's rock and observations you make about it.

Draw what you think the inside of the rock will look like:

Observe It!

Hypothesis:

Tess and her friends will find _____
_____ inside the rock.

How would you observe the inside of the rock?

Again imagine you are with Tess and her friends. How many different things can you think of that you could do to try to open the rock? Which one do you think would work best?

❶ _____
_____

❷ _____
_____

❸ _____
_____

❹ _____
_____

❺ _____
_____

❻ _____
_____

Experiment!

❶ Think about what you have learned about the rock Tess found. What observations can you make about the rock?

_____

_____

_____

❷ Based on your ideas in the *Observe It* section, what things would you do to the rock to learn what is inside it?

_____

_____

_____

❸ What results do you think you might get by doing Step ❷?

_____

_____

_____

❹ Based on the expected results you listed in Step ❸, what conclusion would you draw about the rock Tess found? Would this prove or disprove your hypothesis?

_____

_____

_____

What did you discover?

❶ What kind of rock did Tess find? _____

❷ Was your hypothesis proven or disproven? Why?

_____

_____

❷ How would you compare the rock Tess found to a meteorite?

Geode	Meteorite
_____	_____
_____	_____
_____	_____
_____	_____
_____	_____

❹ Draw a picture of the inside and outside of the rock Tess found.

Discover Roly Polys

If you have ever seen a small bug run across the ground and then roll into a ball, you might have spotted a roly poly.

Roly polys go by different names. Depending on where you live, they can be called butchy-boys, doodle bugs, or pill bugs.

A roly poly is a type of crustacean [crust-AY-shun] that lives on land. A crustacean is an animal that has a shell to protect it, a body divided into segments, and two pairs of antennae. Most crustaceans, like crabs and lobsters, live in rivers, lakes, or oceans, but a few, like the roly poly, live on land.

When a roly poly feels threatened, it will curl into a tight ball to protect itself. They are gentle animals and won't bite or sting. They just like to go about their business, digging in the dirt and looking for moss, bark, or decaying plants to eat. You are most likely to find them in moist places. Look under a rotting log or beneath a pile of leaves.

Photo credits: 1. Franco Folini, 2 & 3. Katya from Moscow, Russia

Tuesday
Observation:
trees are hard.
why are trees so hard?
what makes them
grow tall?

Why don't they
all look the same?
Do trees eat?
Can i get a tree house?
Ask Mom.

Think About It

Tess has lots of questions about trees. How would you answer them?

Take your field notebook outside and look at as many trees as you can. Then sit under a tree and answer the questions.

❶ Why do you think trees are hard?

_____

_____

_____

❷ Why do you think trees grow tall? What do you think makes them able to grow tall?

_____

_____

_____

_____

❸ Why do you think trees don't all look the same?

_____

_____

_____

_____

❹ What do you think is inside a tree?

_____

_____

_____

❺ Write your own questions and answers about trees.

_____

_____

_____

_____

_____

_____

_____

_____

_____

_____

_____

_____

_____

_____

❻ Draw pictures of some tree houses you'd like to live in.

Meteorites

If you have ever looked up into the night sky and watched a shooting star, you might have been observing a meteor passing through the atmosphere. A meteor starts out as an asteroid, which is a rock traveling through space. If an asteroid enters Earth's air, you can see it as it burns up and becomes a meteor, or shooting star. Sometimes small chunks of the meteor survive the journey and hit the ground. When that happens, the rock on the ground is called a meteorite.

Meteorites are made of iron or rock-forming minerals or a mixture of both. Meteorites are the rarest type of rock found on Earth! Tess discovered that it can be hard to tell if a rock is a meteorite or just a regular rock, but sometimes a simple test can help you find out.

Meteorites are usually heavy for their size, may look like the outside has been melted, and are often dark in color. They have an irregular shape, may have thumb print-like pits, and are solid inside. Metal flakes or colorful, sphere-shaped bits of rock can be present in some meteorites. One of the best ways to test if a rock might be a meteorite is to see if it is magnetic!

Photo credits: 1. Meteor, by Navicore;
2 & 3. Meteorites, by H. Raab

Think About It

Think about all the times you have seen the Moon. Write down all the details you can remember about what you have observed at different times. Does the Moon always look the same? Why or why not? Do you think the Moon might be an asteroid? Why or why not? What do you think the Moon is made of? What other questions can you ask about the Moon?

_____

_____

_____

_____

_____

_____

_____

_____

_____

_____

_____

_____

Observe It

Spend several days or weeks carefully observing the Moon. Does the Moon always look the same? Why or why not?

In the space below make drawings of what you see each night. Include all the details you observe. If you have binoculars, use them to see more details. (Use more paper if needed.)

Make a model of the Moon when it is full and looks like a complete circle. Include details. You can use modeling clay, papier mache, or other materials. Paint can be added.

Geodes

When you pick up a rock, it can be fun to think about what might be inside. Most of time when you break a rock open, it looks the same on the inside as on the outside. But sometimes you get lucky, and when you crack the rock open, you find it is hollow and lined with crystals! This type of rock is called a geode.

Geodes can be ball-shaped or oval and look very ordinary on the outside. A geode can have its beginning in an area of volcanic activity or in a deposit of mud that is later turned to rock. Often a bubble of gas is trapped as the rock is formed, resulting in a hollow space in the center. Later, groundwater seeps through the hardened surface of the rock into the hollow interior. Groundwater carries minerals, and over very long periods of time these minerals can form crystals in the hollow center. Geodes come in many different colors, depending on which minerals are in the groundwater. Geodes are found all over the world.

Observe It

Take this notebook and some colored pencils and go for a walk. Look for rocks you think are interesting. Draw each rock and next to the drawing write all the details you observe.

More rocks!

Review your drawings and descriptions of rocks. Compare the rocks and write about the details you observe that are similar and those that are different. What can you conclude about rocks?

_____

_____

_____

_____

_____

_____

_____

_____

_____

_____

_____

_____

_____

_____

_____

_____

_____

Making Geodes

Have you ever wondered what everything is made of? You and all the things around you are made of very, very tiny building blocks called *atoms*. Atoms hook together to make *molecules*.

Molecules can be very simple and made of as few as two atoms, or they can be very complicated and made of thousands of atoms.

Try this little experiment. Put some water in a glass. Put some salt on a spoon and observe it. You can see that the salt is in small pieces called crystals. Now stir the salt into the water.

What happens to the salt? Although the salt seems to disappear, if you taste the water, it tastes salty. So the salt is still there, but where is it? The salt molecules in the salt crystals have broken away from each other and moved in between the water molecules. We say the salt has *dissolved* in the water. Because individual molecules are too small for you to see with your unaided eyes, you can't see either the salt molecules or the water molecules.

In a similar way, rock-forming atoms and molecules can dissolve in water. When water carrying the right kinds of atoms and molecules enters a hollow part of a rock, the atoms and molecules are carried in with the water. When the water leaves, the rock-forming atoms and molecules that are left behind form *crystals*.

sodium   chlorine

A salt molecule has one sodium atom (Na) and one chlorine atom (Cl).

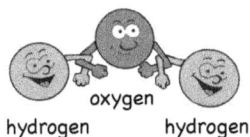

oxygen

hydrogen   hydrogen

A water molecule has one oxygen atom (O) and two hydrogen atoms (H).

Silicate is a rock-forming molecule that has 4 oxygen atoms (red) and one silicon atom (green).

Make Your Own Geodes

Do this experiment to see how crystals form in geodes.

Think About It

❶ Do you think all crystals look alike? Why or why not?

_____

_____

_____

_____

❷ Do you think you could find crystals in other kinds of rocks? Why or why not?

_____

_____

_____

_____

❸ If you were searching for a geode, what characteristics would you look for?

_____

_____

_____

_____

Experiment

Read through the whole experiment, then write your hypothesis (your prediction about what will happen).

Hypothesis _____

_____

_____

Materials you will need

2 or more raw eggs
choose 2 or more solid substances that will dissolve in water:
    table salt
    sugar
    Epsom salts
    baking soda
    borax
    cream of tartar
water
food coloring
saucepan
egg carton or pan for mini muffins
wax paper
several cups (heatproof)
spoons
bowl

Experimental Steps

❶ Carefully crack each egg near the smaller end and remove the end piece of eggshell. The main part of the eggshell needs to stay whole so it can hold liquid. Pour the egg whites and yolks into a bowl.

❷ Eggs have a membrane (skin) on the inside of the shell, and this membrane needs to be removed. Carefully pour hot water into the eggshell. This will cook the membrane. Pour the water out of the eggshell and use your fingers to remove the membrane. If part of the membrane is left behind, it can get moldy and make your crystals turn black, so make sure that all of it is removed

❸ Put the cleaned eggshells open end up in an egg carton lined with wax paper or in a mini muffin pan. You may need to prop them up so they won't tip when you fill them with water.

Have your teacher help you with the next steps.

❹ Heat water to boiling in a saucepan. You will need about 1/2 cup (118 ml) of water for each egg.

❺ Pour about 1/2 cup of boiling water into a cup. Stir in 1/4 cup of one of the solid substances until it dissolves in the water. Then add more of the solid a little bit at a time until no more will dissolve.

❻ Stir in some food coloring.

❼ Carefully pour the colored water into an eggshell until it is full but not overflowing.

❽ Repeat Steps ❺-❼ for each eggshell.

❾ Put the filled eggshells in a safe place. Check them every day for several days to see what happens.

❿ Beat the egg whites and egg yolks and make an omelet.

Conclusion

What were the results of your experiment? How would explain the results? Was your hypothesis proven or disproven?

_____

_____

_____

_____

_____

_____

_____

Why?

In this experiment you created supersaturated solutions. A solution is a liquid that has a solid dissolved in it. In a solution the molecules of a solid have moved in between the molecules of the liquid. A supersaturated solution has all the spaces in between the liquid molecules filled with molecules of the solid. A liquid that is hot can hold more molecules of a solid, but when it cools, it can't hold as many. As the liquid cools, molecules of the solid settle out and accumulate to form crystals.

The crystals in geodes form in a similar way, with dissolved minerals settling out and accumulating slowly. Different rock-forming minerals will dissolve in water, and the type of mineral that forms a crystal will determine the color and shape of the crystal. The crystals in geodes form very slowly over a very long period of time.

Falling Objects

Think about it

❶ What do you think would happen if you dropped an apple and a marble at the same time from the same height? Why?

_____

_____

_____

_____

❷ What do you think would happen if you dropped an apple and a marble at the same time from the same height but the height was higher or lower? Why?

_____

_____

_____

_____

❸ What do you think would happen if you dropped an apple and a feather at the same time from the same height? Why?

_____

_____

_____

_____

Drop It!

Galileo was a famous Italian scientist who was very curious about how things move. It is said that he dropped objects off the Leaning Tower of Pisa to observe how things fall. Try Galileo's experiment to find out what happens when different objects fall at the same time.

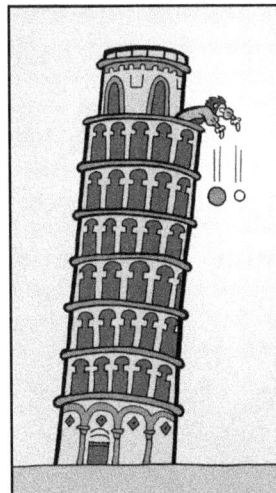

Materials you will need

two or more pairs of balls of different weight

    e.g.: basketball & tennis ball

           ping-pong ball & tennis ball

           ping-pong ball & baseball

Experiment

❶ Choose two balls of different types.

❷ From a standing position, drop the two balls simultaneously from chest height. Observe whether one ball strikes the ground first or whether both balls strike the ground at the same time. Record your results below.

❸ Repeat Steps ❶-❷ one or more times using two different balls each time.

Results

_____

_____

_____

_____

_____

_____

Conclusions

Many early scientists believed that a heavier object will strike the ground before a lighter object. Galileo's experiment shows that objects of unequal weight will hit the ground simultaneously. This happens because Earth's gravitational force is the same for light objects as it is for heavy objects.

If you did not observe both objects striking the ground at the same time, think of some ways that you might change your method of observation to be more accurate. Write your ideas below and then repeat the experiment one or more times, using your ideas.

_____

_____

_____

_____

_____

_____

_____

_____

_____

_____

_____

??? WHAT DOES IT MEAN ???

asteroid [AS-ter-oyd] • a rock that is moving through space

atom [A-tum] • a basic building block of a substance

comet [KAH-met] • a ball of ice and dirt that is moving through space

conclusion [kun-CLUE-zhun] • the last step in the scientific method; the summary of the results or outcome of an experiment

crustacean [crust-AY-shun] • an animal that has a shell to protect it, a body divided into segments, and two pairs of antennae; examples are roly polys, lobsters, and crabs

crystal [KRIS-tuhl] • a type of mineral that has an organized internal structure of atoms; can have smooth, flat sides; can be shiny

dissolve [di-ZAHLV] • when the molecules in a solid break apart from each other and go in between the molecules of a liquidw

experiment [iks-PER-uh-ment] • a step-by-step series of actions used to prove or disprove a hypothesis or idea

force • the power that changes the position, shape, or speed of an object

Galileo [ga-luh-LAY-oh] • (1564-1642 CE) a famous Italian scientist who studied how things move

geode [JEE-ode] • a rock that has a hollow interior that is lined with crystals

hypothesis [hi-PAH-thuh-sis] • a statement about something that has been observed; a prediction or guess about what the results of an experiment will be

igneous rock [IG-nee-us rock] • rock that has its beginning as melted rock deep within the Earth; igneous rocks can be smooth or rough, shiny or not shiny, and may contain crystals

law of motion • a scientific statement that describes how objects move and what makes them move

metamorphic rock [met-uh-MORE-fik rock] • a type of rock that is made from other rocks that long ago sank deep into Earth and were melted; a metamorphic rock may or may not have layers

meteor [MEE-tee-or] • an asteroid that has entered Earth's air and is burning up; also called a shooting star

meteorite [MEE-tee-or-ite] • a rock that results from a meteor reaching the surface of Earth

mineral [MIN-rul] • a naturally occurring, solid material that has an organized internal structure of atoms; found in rocks

molecule {MAH-luh-kyool] • two or more atoms joined together

observation [ob-sur-VAY-shun] • the act of looking at something carefully and in detail

quartz [kwortz] • a common mineral

rock-forming mineral [MIN-rul] • a mineral commonly found in rocks

scientific method • a series of steps used by scientists to make discoveries

sedimentary rock [sed-uh-MENT-uh-ree rock] • a type of rock that is made from bits of rock that have been pressed together

shooting star • a meteor

Make Your Own Glossary!

When you learn a new word, add the word and its definition to the glossary.

_____

_____

_____

_____

_____

_____

_____

_____

_____

_____

_____

_____

_____

More Glossary!

Field Notebook!

This section is waiting for you to record your observations, thoughts, questions, ideas, and discoveries!

www.ingramcontent.com/pod-product-compliance
Lightning Source LLC
Chambersburg PA
CBHW051354200326
41521CB00014B/2570